Motion in SPORTS

Written by

David Wiseman

Designed by

Robin Fight

Sports have played an important role in the world for thousands of years! In fact, the earliest sporting games—such as running, wrestling, and throwing objects—appear to have been created to practice basic survival skills. Today, sports can bless your life in many ways. Apart from providing the health benefits of exercise, sports can also teach you important lessons about teamwork, kindness, and leadership.

Did you know sports can also help you learn more about science? Most people probably aren't thinking about the scientific method while cheering for their favorite team, but God's natural laws are a big part of all sporting activities. In this book you will learn about fascinating links between a specific branch of science called *physics* and a wide range of amazing sports. As you read, try to think of new ways to use the laws of force and motion to enjoy the sports you love even more.

Before we start, let's take a look at some key terms you will see repeated in this book. You might want to refer back to them as you read. Understanding these concepts will help you find more success in any sport!

FORCE: A push or pull that causes an object to have a change in movement or direction

MOTION: The act of moving an object to a different place or position

INERTIA: The tendency of an object to stay either at rest or in motion until it is acted upon by another force

FRICTION: The resistance of motion when two objects rub against one another

SPEED: How far an object travels in a certain amount of time

VELOCITY: The speed and direction of an object

MASS: The measurement of how much matter is in an object

GRAVITY: A force of attraction between two things

KINETIC ENERGY: The energy an object has due to its motion

POTENTIAL ENERGY: The energy stored in an object due to its position, tension, or charge

BASEBALL

Going to the ballpark means more than simply watching a sporting event. It is a full experience with tastes, smells, sights, and sounds that are unique to the game. Beyond its cultural importance, baseball is also a competition that relies as much on the principles of physics as it does on strength and athleticism. When it comes to combining sports and the laws of motion, baseball definitely hits a home run!

Pitching

You may think that pitching is all about throwing the ball as hard as you can. Speed is definitely a big part of the game, but spin might be even more important. As a ball rotates, air passes more quickly in the direction of the spin to reduce air pressure, making the ball move in that direction. This is known as the Magnus Effect.

QUICK TIP

Momentum (the strength an object has when it is moving) = mass x velocity ($p = mv$). This means that finding the ideal balance between the weight of your bat and the speed of your swing will maximize the impact of your hits.

Take a look at how some basic pitches use different spins and speeds to keep a batter guessing.

	FASTBALL	SINKER	CURVEBALL	KNUCKLEBALL
SPIN	Fast backspin	Sidespin to minimize backspin	Harder sidespin	Minimal spin
MOVEMENT	Moves straight or slightly upward	Moves downward at end of pitch	Moves left or right and downward	Unpredictable movement
PRO SPEED	136–160 km/h (85–100 mph)	136–157 km/h (85–98 mph)	104–136 km/h (65–85 mph)	88–112 km/h (55–70 mph)

Bat/Ball Collision

There are few sounds as exciting for a sports fan as the contact of a bat with a speeding baseball. It also offers an amazing example of physics in action! Both the bat and the ball have kinetic energy due to their fast movement, but the bat has more mass, so when they collide, momentum moves from the bat to the ball. As they meet, the contact actually deforms the size of the ball momentarily before sending it soaring through the air in a new direction.

Have you ever watched gymnastics and wondered how it is possible for the competitors to spring into the air with such power? If so, you are not alone, as the sport has amazed fans since its earliest days in ancient Greece. Gymnastics requires a combination of strength, speed, agility, and an understanding of the laws of physics to rise to the challenges of competition and reach the medal podium.

Center of Gravity

Center of gravity is the point in an object where its weight is balanced on all sides. Finding the center point of the body's mass is needed to keep balance when moving. This can be especially challenging during events like the balance beam, where competitors have to jump, twist, and flip on a beam only 10.16 cm (4 in) wide! Shifting the body even slightly starts a battle between the force of gravity and the strength of the gymnast, and if the center of gravity gets too far off-center with the beam, then gravity will ultimately win.

GYMNASTICS

Vault

Vaulting is one of the most exciting events in gymnastics! It is also packed with physics principles that start with a full sprint and hopefully end with a perfect landing.

	APPROACH	SPRINGBOARD	VAULT	FLIGHT	LANDING
ACTIVITY	The gymnast runs 25 m (82 ft) or less as fast as possible toward the vault.	The gymnast does a low jump or handspring onto a springboard in front of the vault.	The gymnast places his or her hands on the vault table and pushes the body upward.	The gymnast performs a variety of flips and twists in the air.	The gymnast lands on his or her feet with as little movement as possible.
SELECT FORCES	The speed of the run gathers kinetic energy used to increase the force of contact with the springboard and vault.	Downward force causes the springs to compress, becoming elastic energy that pushes back against the gymnast.	Hands create force against the vault and help align the body's center of gravity for more lift and rotation.	Rotating around axis points on the body is possible due to a spinning force called torque.	Legs are used as shocks to absorb the force when the gymnast hits the landing pad.

QUICK TIP

See if you can feel your center of gravity shift as you safely do a variety of twists and jumps.

ICE HOCKEY

Ice hockey is one of the fastest and most physical sports in the world. You must have incredible strength and agility to skate, pass, shoot, and hopefully score more goals than the opposing team. The sport also uses a lot of science, especially since it is played entirely on ice. From sliding to shooting at the goal to colliding with other players, hockey is a perfect sport for studying force and motion at work.

Skating

The two most basic motions of skating are pushing and sliding. Hockey skates have a blade that is typically 2.9 mm (0.11 in) wide. This thin blade helps to lower the friction with the slick ice. Apart from allowing hockey players to accelerate to top speeds of 38.62 km/h (24 mph), the ice is also an ideal surface for stopping quickly and moving in different directions by carving the sharp blade into the frozen surface. Players can even skate backward, moving in an "S" pattern in order to avoid turning their backs on their opponents.

Slap Shot

The slap shot uses many types of force and energy to shoot a 170-g (6-oz) puck at a speed of over 160 km/h (100 mph) toward the goal. Check out some of the steps that make this possible.

QUICK TIP

Any time you step onto the ice, be sure to wear protective gear that fits properly to help avoid injuries.

1. The hockey player lifts the stick behind the body. The raised stick holds potential energy due to its position, which then converts to kinetic energy through a powerful swing.

2. The stick first hits the ice about 8–13 cm (3–5 in) before the puck. The hockey player actually bends the stick at nearly a 30-degree angle. Elastic energy is stored in the bent stick before it springs forward.

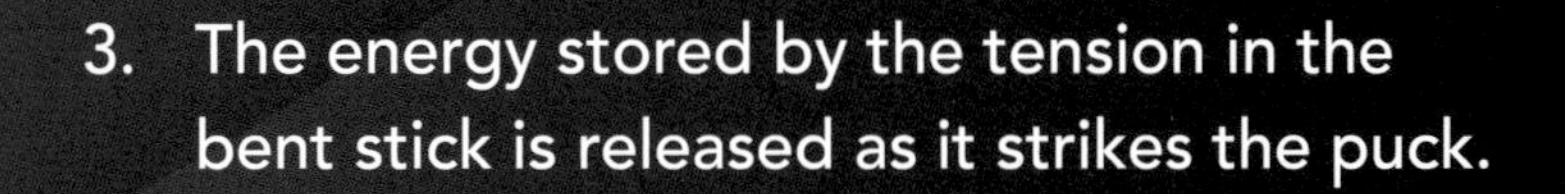

3. The energy stored by the tension in the bent stick is released as it strikes the puck.

4. The hockey player also shifts body weight into the swing to create even more force.

5. The curve of the stick's blade helps lift the puck slightly off the ice so it wobbles less and reaches maximum speed as it flies toward the goal.

With the help of a flexible pole, a pole vaulter can fly into the air to clear a bar as high as a two-story house. At first glance, this sport might seem easy to understand, but there are a lot of complex physics happening behind the scenes. Athletes must learn about velocity, body angles, and a variety of energy types. Pole vaulting has raised the bar on the science of motion!

Energy Transfers

Several types of energy are at work during a pole vault, and as each action is performed, this energy is transferred from one thing to another. The most successful pole vaulters are those who fine-tune these movements so everything works together. Let's take a look at how this transfer of energy works.

KINETIC ENERGY:

Kinetic energy comes from motion. This is why the pole vaulter needs to sprint down a runway at least 40 m (43.75 yd) long at top speed. The greater the speed, the more energy is available to shift the forward motion to an upward direction for the big leap.

ELASTIC POTENTIAL ENERGY:

At the end of the run, the vaulter must place the end of the pole in a small box that secures it to the ground. At that moment, energy transfers to the pole. Similar to a spring, the pole stores what is called elastic energy as it bends. Then, the athlete must change body position to point upward with the feet above the head.

GRAVITATIONAL POTENTIAL ENERGY:

As the pole straightens out again, the elastic energy is released as gravitational potential energy. This is simply the amount of force gravity has on an object before it returns to the earth. In other words, the energy stored in the pole goes to the athlete, who once again must move the body strategically to clear the bar before safely landing on a padded mat.

QUICK TIP

Before you ever take a leap, practice short sprints with your body as tall as possible and your knees high.

The art of figure skating has advanced a lot since the first ice skates were invented using animal bone thousands of years ago. Today, figure skating combines power with elegance to delight fans through gliding grace and dizzying spins. Beyond the artistry and athleticism, it uses the study of force and motion to help competitors reach their goals.

Jumps

You don't have to know the definition of a triple Salchow to appreciate the power and precision of figure skating jumps. Competition jumps are split into two basic categories—edge jumps, where the skater launches using the blade, and toe jumps, which are performed with the help of the toe pick at the front of the skate. There are six official figure skating jumps, each involving one-half to four rotations, but let's look at some common elements of physics they all share.

LAUNCH

The launch must find a balance between height and rotation. If a jump is too high, there won't be enough momentum to complete the spin, but too little height means not enough time to finish the rotation.

JUMP

The skater generally has less than a second to complete the jump. He or she starts with arms extended and then pulls toward the body to rotate faster. A quadruple spin requires a speed of around 340 rotations per minute.

LANDING

When landing, the skater must absorb the energy of the height and spin of the jump at the same time. Scientific studies have shown that the force of impact can be five to eight times the body weight of the skater!

Moment of Inertia

Have you ever wondered how it is possible for figure skaters to spin so fast? The answer has to do with a scientific concept called *moment of inertia*. This means that the closer the mass of a spinning object is to its center axis, the more easily it will rotate. This is why figure skaters start spins with their arms extended and then bring them in close to their bodies. It decreases the moment of inertia and allows them to spin so fast they look like blurs!

FIGURE SKATING

QUICK TIP

Try spinning in an office chair with outstretched arms, and then suddenly pull your arms toward the center of your body to increase speed.

SOCCER

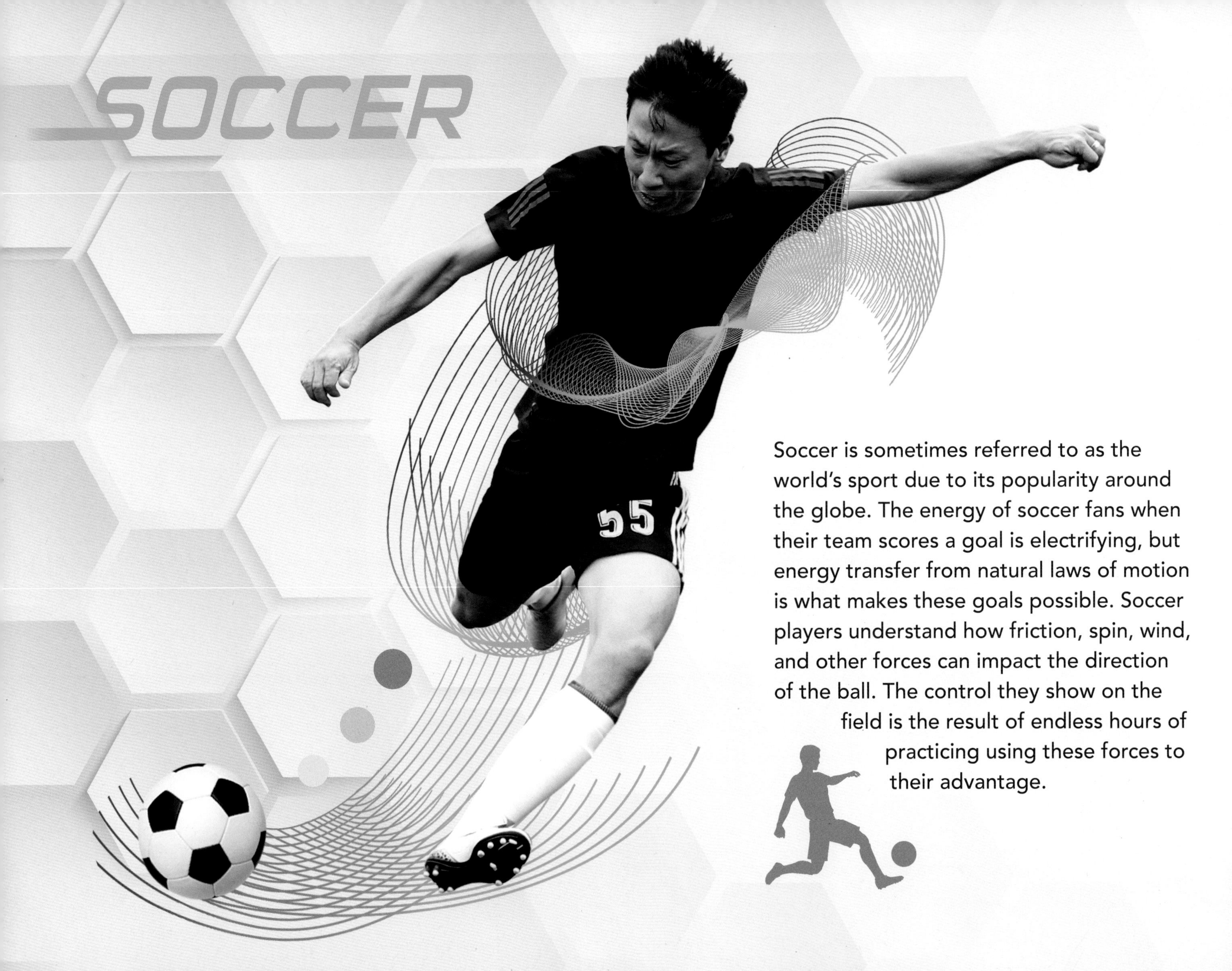

Soccer is sometimes referred to as the world's sport due to its popularity around the globe. The energy of soccer fans when their team scores a goal is electrifying, but energy transfer from natural laws of motion is what makes these goals possible. Soccer players understand how friction, spin, wind, and other forces can impact the direction of the ball. The control they show on the field is the result of endless hours of practicing using these forces to their advantage.

Bending a Kick

Sometimes a soccer ball seems to mysteriously curve in the air as it flies into the net to score a goal. This curve actually isn't a mystery at all. It's due to the Magnus Effect. This means a spinning object with forward velocity will move in the direction of its spin due to lower air pressure on that side. If you want a soccer ball to bend left, kick it slightly on the right side. To go right, kick it on the left. The force of the kick, how high the ball is kicked, the direction of the wind, and other factors can also change the ball's curve. It takes a lot of practice to get everything just right, but experimenting with the Magnus Effect can lead to more goals!

Surface Friction

The surface a soccer match is played on impacts the way the players and the ball move. Most of the time, matches are played on grass fields. Friction between the ball and the grass slows the ball down. Players, on the other hand, sometimes slip on the slick grass, which is why they wear shoes with spikes to grip the ground as they run. Playing on dirt or a solid surface like cement changes the speed of the game completely since there is less friction with the ball, causing it to move much faster.

QUICK TIP

Kicking a soccer ball with the side of your foot gives you more control for passing, but kicking it on the laces of your shoe will give you the most power.

Archery is an ancient human activity. It has been used for hunting, combat, competitive sport, and more. Shooting an arrow with accuracy requires strength, a good eye, and a lot of physics. Velocity is a combination of speed and direction, so it takes both the force of the bow and the aim of the archer for an arrow to strike the target!

Bow Energy

The name archery comes from the Latin word *arcus*, which means bow. The arrow is what actually hits the target, but it's the bow tension that generates most of the energy. The force of a tightly pulled string is known as draw weight and is stored as potential energy. The draw weight of a bow can range from 6.80 kg (15 lb) of pressure for a children's bow to as much as 90.72 kg (200 lb) of pressure for what is called a war bow! The poundage of a bow changes depending on how far the bowstring is pulled back. The string stays the same size, but the bow changes shape like a giant spring. This is called elastic energy. Upon release, some of that energy moves to the bow as it retakes its shape, but most of the force goes to the arrow, which can fly through the air at speeds of more than 321.87 km/h (200 mph)!

QUICK TIP

If you want to begin participating in archery, first find an expert to help you get the right bow fit, work on form, and learn safety rules.

ARCHERY

Archer's Paradox

On many bows the arrow sits on a shelf at one side of the bow. This means the arrow points slightly sideways when at rest, but when it's shot, it moves to the center. This is called the Archer's Paradox. What seems like a mystery is actually physics in action. The sudden release of stored energy onto the arrow forces it to curve around the bow's limb and align with the string's centered position. It's important for the arrow to be stiff enough to remain stable yet flexible enough to move around the bow limb and retake its shape to hopefully hit a bullseye!

Curling is a fascinating sport where players on two competing teams of four members each push heavy granite stones on a strip of ice over 45 m (49.21 yd) long toward a target called the house. After one team member releases the stone, the others use special brushes on the ice ahead of the stone to help it move faster, slower, or curve. Teams score points for each stone inside or touching the house that is closer to the center than all of the opposing team's stones. This sport is full of strategy to use the laws of physics to maximum advantage. In fact, thinking time is actually built into matches for teams to figure out their next move!

CURLING

Sweeping

One of the more intriguing parts of the game for many spectators is sweeping. Small water droplets called pebbles are frozen on top of the ice to limit the curl or curve of the stone. The quick sweeping motion smooths the path for the stone and slightly raises the temperature, making the ice slicker and lowering friction. Sweeping helps the stone travel in the desired path to place it in the perfect spot to score or knock another team's stone out of the house.

Curling Yells

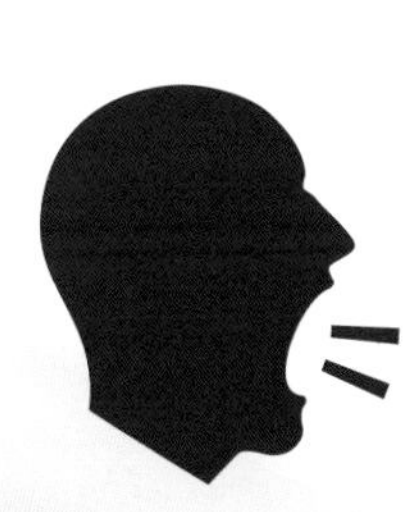

Curling is a sport with a long tradition of honor, respect, and civility based on a code of conduct called the Spirit of Curling. That said, if you watch curling events, you might hear a lot of yelling. This has nothing to do with being rude. As team members sweep the ice to change the path of the stone, they shout instructions to communicate with each other. Check out some of the most common curling shouts and the actions that go with them.

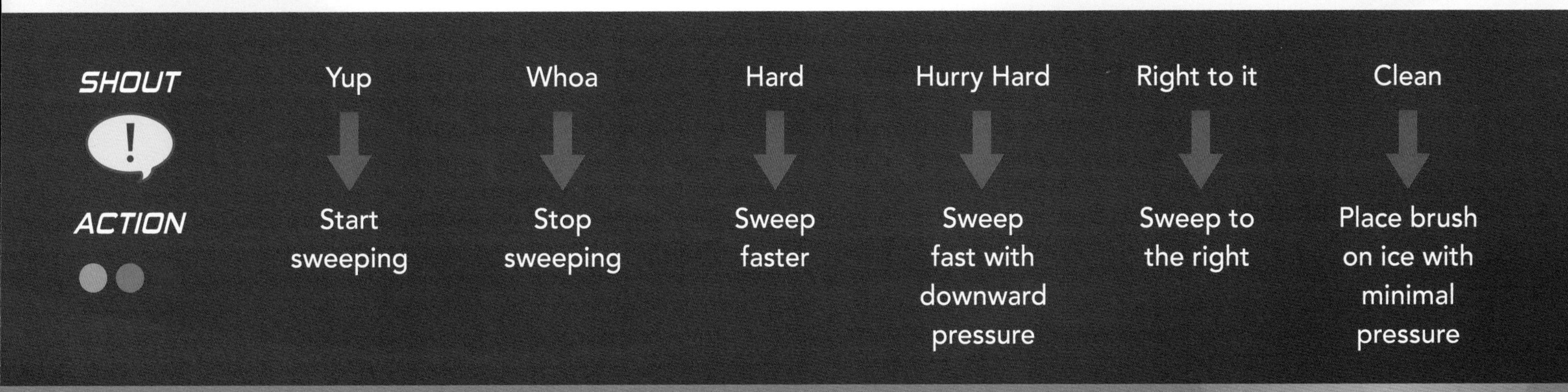

QUICK TIP

Practice reacting to quick curling shouts from a friend using a regular cleaning broom.

GOLF

The concept of golf is simple: try to hit a small ball into a faraway hole in as few shots as possible. Those who try it for the first time learn very quickly that it is more complex than it looks. Golfers must understand force, motion, angles, and more to be successful. Learning the concepts below will help you get to the green, where the hole resides, and stay out of the rough or other obstacles along the way.

Golf Ball

In the early days of golf, players noticed that their older balls with dings and scratches went farther distances than the smooth new ones. If you look at golf balls today, they are intentionally made with small dents called dimples. As air passes over a smooth ball, the air does not stay close to the ball. This creates a force called drag that slows the ball down. The dimples in the ball interrupt the airflow and make it attach more closely to the ball's surface. This reduces drag so much that the ball can literally go twice as far!

Double Pendulum

A golf swing uses the force of two pendulum motions. The first pendulum uses the shoulders as the axis to swing the club above the head and then move it toward the ball. This converts potential energy to kinetic energy. The second pendulum uses the wrists to swing the club forward again right before it hits the ball. The contact only lasts a fraction of a second, but this double pendulum creates a force of impact up to 907.18 kg (2,000 lb)!

Friction

Friction with a variety of surfaces at a golf course can present challenges for any golfer. Short grass can slow a ball down, but the ball still bounces and rolls. The rough on the sides has long grass, meaning less movement. Bunkers, or valleys filled with sand, are especially tricky because a ball can actually get stuck! Once you get to the putting green, close to the hole, the grass is very short, so the ball rolls fast. The putting green also often has slopes, so you might see a player get down on the ground to look at the angles before putting.

QUICK TIP

Try observing the swings of other golfers then practicing the motion of your swing. Practice using small movements to help you notice the pendulum motions used in a great swing. See if making slight changes helps you increase your swing's force.

Modern-day basketball was invented in 1891 when Dr. James Naismith nailed up some peach baskets in a gym. It has grown to become one of the most popular sports in the world. From dribbling to passing to jump shots, basketball players use a wide range of motions and forces to find success on the court. If you want to experience a quick-paced sport packed with physics, then basketball is a slam dunk!

Dribbling

Dribbling a basketball is one of the best examples of Newton's laws of motion. The ball is at rest until a player decides to push it toward the floor. Velocity and mass create the force of the bounce, and then the ground pushes back with an equal and opposite force to return the ball to the player's hand. Higher air pressure within the ball will help the ball bounce more because the air inside also pushes against the impact on the court floor. All these forces are in play almost at the same time, so a player must practice a lot to keep the dribble low, fast, and away from defenders.

Shooting

There are many ways to shoot a basket, but backspin is one thing that almost all great shooters have in common. Backspin is created with the hand and wrist when the ball is released. This helps the ball roll toward the net if it hits the rim or backboard. Shooters also need to practice the angle of their shots. Having more arch increases the chance of making a basket because the ball travels downward at the end of its motion instead of going straight at the basket.

QUICK TIP

Squaring up your shoulders perpendicularly with an imaginary line from you to the rim can help straighten your shot before you release the ball.

TENNIS

Tennis is a fast-paced sport played on a variety of surfaces. Expert players have to make split-second decisions to react to the direction of the ball and hit it back with the right amount of spin and velocity. They have to be in great shape to quickly run back and forth on the court, but they also need to be in top condition mentally to understand and apply laws of physics.

Court Surfaces

Tennis can be played on many different surfaces. Check out some of the differences below.

GRASS	CLAY	HARD COURTS
Fast ball speed	Slower ball speed	Medium ball speed
Low bounce	High bounce	Highest bounce

Serving

The serve begins when a player tosses the ball in the air before striking it. It's important to hit the ball at its highest point because this is where the gravitational potential energy is at its peak. At impact, it is not just the force of motion from the swing that sends the ball flying to the other side of the net. It is also the elastic energy from the racket's strings. They act like springs to absorb the contact and push it back to the ball. Finally, the downward angle of the strike means that gravity is also working in the server's favor to increase speed. The combination of all these forces allows the most powerful serves to reach over 241.40 km/h (150 mph)!

Spin

Tennis players use spin and angles to control the flight of the ball. Topspin makes the ball move downward. This type of spin is created by angling the racket downward but swinging upward to make the ball spin forward. Players can hit the ball harder using topspin because its downward motion helps it stay on the court. Backspin (when the ball spins backward) also has its advantages. It keeps the ball higher in the air, making it harder for the opposing player to return a shot with as much force.

QUICK TIP

The "sweet spot" on a tennis racket is where the strings hit the ball with the most power. Try holding a racket on your lap and dropping the ball on it to see where it bounces highest.

The ball leaves your hand. It's flying toward the pins. *Crash!* You got a strike! Bowling is an exciting sport that is more challenging than it might seem. It's not always the strongest throw that wins. In fact, throwing too hard or too straight can actually earn you a lower score. The best bowlers understand the importance of spin, velocity, friction, and angles in knocking down as many pins as possible.

Path of a Bowling Ball

Many factors determine the path of a bowling ball. The weight of the ball and its velocity are important, but so are the lanes themselves, since oils used to protect their surface can impact the path of your ball. Another important aspect of the game is the release point of the bowling ball. Any bounce in the ball will slow it down, so bowlers try to release the ball as close to the ground as possible. The ideal path of the ball curves toward the center of the pins instead of hitting them straight on. This is called a hook, and it helps distribute the impact to all of the pins more evenly rather than just hitting a few of them with all the force directly.

QUICK TIP

When starting out with bowling, don't try to throw the ball too hard. Work on your release and the path of the ball first before increasing speed.

Bowling Pins

Bowling pins carry most of their weight at the bottom. They are also rounded, so they roll when hit. The combination of their low center of gravity and rounded edges means that bowlers can hit the pins strategically to have one pin knock down others. This is especially important in a shot called a 7-10 split where there is one pin left on each side of the lane after the first attempt. The goal is to throw the ball at the inside of the pin opposite your bowling hand with the hope that it will bounce with enough force off the wall and back to the other side of the lane to hit the other pin. Don't worry if you aren't successful because even the pros only make this shot 0.7% of the time!

BOWLING

LONG JUMP

Did you know that the best long jumpers in the world can leap nearly 9 m (9.84 yd)? That's like jumping over two cars laid end to end! Long jumpers have amazing strength. They also use timing, technique, and science to perfect their form, maximize their time in the air, and increase distance as much as possible. Every inch counts in this sport, so competitors take full advantage of the laws of physics to stretch the limits of their jumps.

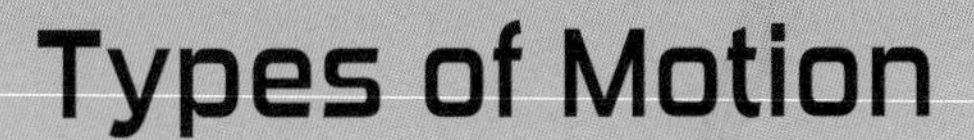

Types of Motion

Success in the long jump depends upon two directions of motion—one forward and the other upward. The velocity from the run before the leap is what carries the athlete forward. Some kinetic energy transfers into the upward movement, but most of it comes from the jumper springing off a launch board at the end of the runway. The best angle for projecting objects the farthest distance is usually 45 degrees, but experts have found that the ideal launch angle for most long jumpers is closer to 22 degrees.

Four Stages

Take a look at the four basic parts of the long jump with instructions for each action and select forces at play during each stage.

	RUN	LAUNCH	FLIGHT	LANDING
ACTION	Long jumpers sprint as quickly as possible toward the launchpad, reaching maximum speed at the end of around 40 m (43.74 yd). The last two steps should be the longest strides.	The foot is pushed down flat on the launch board with as much stiffness and force as possible. The free leg, head, and arms swing forward at the same time to increase upward force.	There are many flight techniques, but in the end, the legs are stretched forward as far as possible to increase the distance between the launch board and the spot of landing.	Long jumpers drive their heels into the sand first and then use their knees as shocks to try to push their bodies over their feet to fall forward as much as possible.
SELECT FORCES 	The velocity of a run transfers its kinetic energy into both the forward and upward motions of the jump.	The locked leg acts like a tight spring to push downward on the board. This causes an equal and opposite reaction upward for maximum lift.	Kicking the legs out not only stretches the body to increase distance, but the force can also push the long jumper forward.	Strong forces push the body backward at the landing, so it takes a lot of muscle strength to change momentum to rotate forward.

QUICK TIP

Lowering your head to see the launch line can slow you down, so practice the timing of your approach to hit the board without looking.

Surf's up! That means it's time to hit the beach for one of the most amazing displays of physics in sports. Catching waves involves defying gravity through an upward force on the surfer's board called buoyancy. Surfing is a challenging mix of balance, strength, and multiple laws of motion acting together to allow competitors to glide on water with power and style.

SURFING

Gravity vs. Buoyancy

Surfing can be very competitive, but one of the most important battles doesn't involve the athletes. Gravity and buoyancy compete with each other to determine whether a surfer will sink or float. As gravity pulls the surfer down, the buoyant force of the water pushes upward. Density, or the amount of matter within an object, plays an important role here. Early surfboards were very heavy and long. This made them sink and not move easily, but now they are much lighter, less dense, and more buoyant. Surfers add weight when they get on their boards, so they typically sink in still water, but the thrust of the waves helps to push the surfers forward on top of the water. Knowing how to strategically shift their weight helps surfers change directions and get where they want to go. If surfers forget to keep their center of gravity low, they are more likely to wipe out, or fall off their boards.

Wave Energy

When you look at waves at the beach, it might seem like the water is traveling from far away to finally reach the shore, but that isn't accurate. Some water does move with the waves, but it doesn't travel far. What does travel is the energy that passes through the water. First, it creates ripples, which then swell into larger waves. Surface waves, caused mostly by wind creating friction with the water, are the most common type. As waves pass over shallower water, the water from the lower part of the wave pushes upward, making the top of the wave curl forward. This creates the tube, or barrel, that surfers love. For a sport like surfing that depends upon waves to ride, understanding how they form, curl, and move can help you to eventually hang ten or ride the barrel of a wave.

QUICK TIP

As you first begin surfing, try using a larger, soft-top board. They are very buoyant, more comfortable when paddling, and extra stable to help you keep balanced.

In this book we have only looked at a handful of sports that involve laws of force and motion. Can you apply what you have learned to other sports? The actual word "physics" comes from the Greek word *φυσική*, which means "natural." Understanding how things move and respond to forces in nature can help you use these same laws of physics to enhance the way you interact with the world around you. This can also help you become a better athlete. Every time you play outside or practice a sport, always remember there are fascinating physics principles at work!